I0797710

Natural Resources

SOIL

Amy C. Rea

DiscoverRoo
An Imprint of Pop!
popbooksonline.com

abdobooks.com

Published by Pop!, a division of ABDO, PO Box 398166, Minneapolis, Minnesota 55439.

Printed in the United States of America, North Mankato, Minnesota.

102019
012020

THIS BOOK CONTAINS RECYCLED MATERIALS

Cover Photos: iStockphoto (grass, soil)
Interior Photos: iStockphoto, 1 (grass), 1 (soil), 5, 8, 9, 13, 19, 23, 26, 28 (bottom), 30; Shutterstock Images, 6, 7, 11, 12, 14–15, 17, 18, 20–21, 24–25, 27, 28 (top), 29 (top), 29 (bottom), 31

Editor: Sophie Geister-Jones
Series Designer: Jake Slavik

Library of Congress Control Number: 2019942465

Publisher's Cataloging-in-Publication Data

Names: Rea, Amy C., author.

Title: Soil / by Amy C. Rea

Description: Minneapolis, Minnesota : Pop!, 2020 | Series: Natural resources | Includes online resources and index.

Identifiers: ISBN 9781532165870 (lib. bdg.) | ISBN 9781532167195 (ebook)

Subjects: LCSH: Soils--Juvenile literature. | Dirt--Juvenile literature. | Natural resources--Juvenile literature. | Environment--Juvenile literature. | Ecology--Juvenile literature.

Classification: DDC 577.57--dc23

WELCOME TO DiscoverRoo!

Pop open this book and you'll find QR codes loaded with information, so you can learn even more!

Scan this code* and others like it while you read, or visit the website below to make this book pop!

popbooksonline.com/soil

*Scanning QR codes requires a web-enabled smart device with a QR code reader app and a camera.

TABLE OF

CONTENTS

CHAPTER 1
IN THE GROUND

A worm tunnels through the dirt. It passes a seed. The seed has begun to grow. Roots reach down into the soil. They take in **nutrients** from the dirt. The plant uses these nutrients to grow.

WATCH A VIDEO HERE!

Earthworms bring nutrients from deeper in the soil to the surface.

Soil is more than just dirt. Soil is made of many things. It contains small pieces of rocks and **minerals**. It also

Erosion breaks tiny pieces of rocks off of the bigger rock. These tiny pieces become part of the soil.

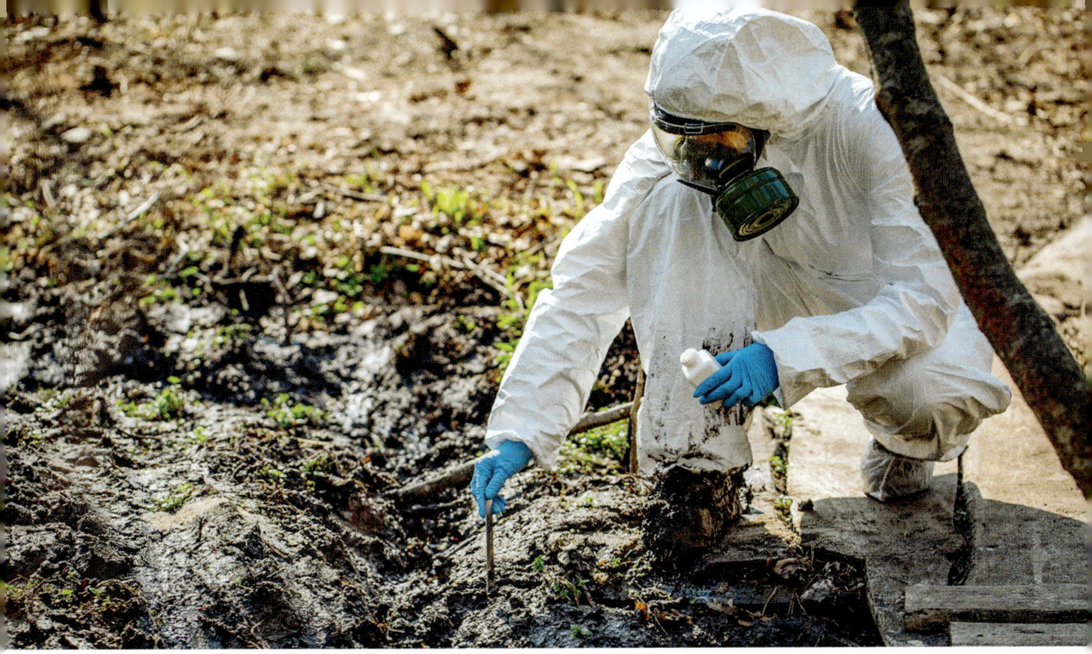

Scientists study soil and the organisms in it.

contains water and air. Parts of dead plants and **organisms** are found in soil. And it is home to many living things.

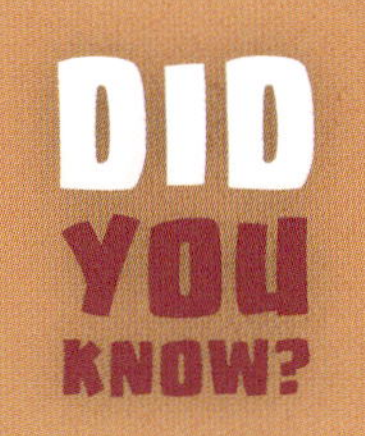

DID YOU KNOW? **A tablespoon of soil has more organisms in it than there are people on Earth.**

Soil forms over many years. Bits of rock break off from Earth's crust. They combine with living and dead organisms. Everything is mixed together. All these factors affect the soil's color and texture.

High amounts of iron give soil a red color.

SOIL LAYERS

Soil is made up of different layers. Some types of soil have as many as six layers. Most soils have three. Topsoil is at or near the surface. Subsoil comes below it. Many minerals seep down into this layer. Parent material is at the bottom. This layer is part of Earth's crust.

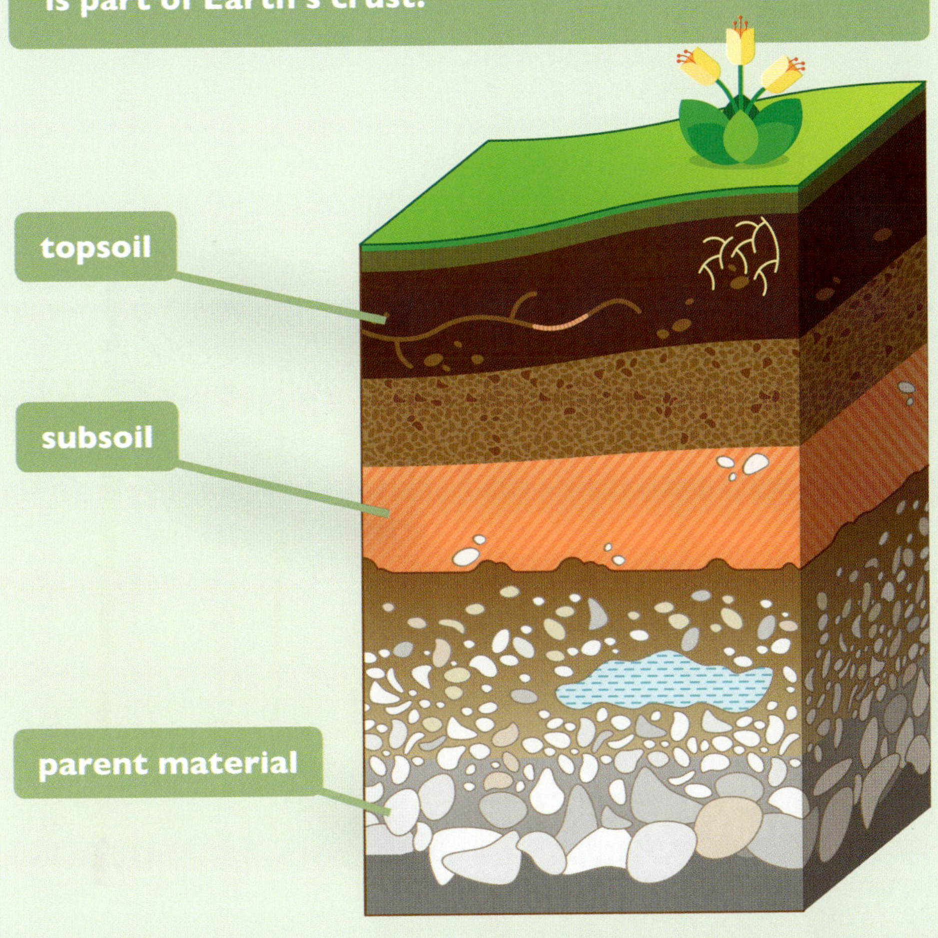

CHAPTER 2

SOIL DOES SO MUCH

Soil is found all over the planet, even at the bottom of the ocean. There are many different kinds of soil. All are important to life on Earth. Many plants and animals depend on soil.

LEARN MORE HERE!

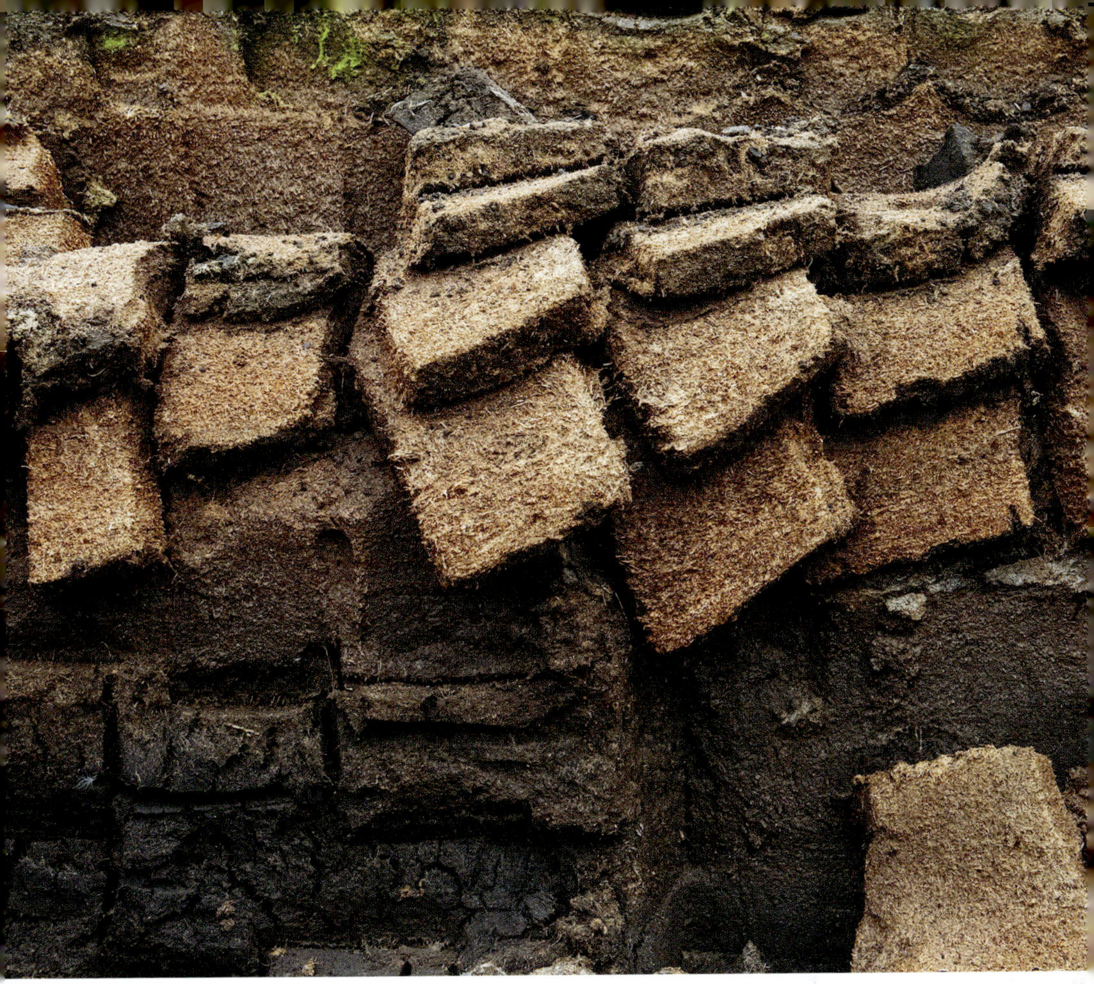

Peat is a dark, wet soil that can be burned when dried out.

DID YOU KNOW?

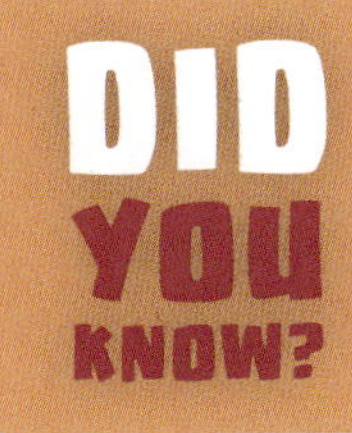

There are 70,000 types of soil in the United States.

Soil helps hold things together.

It holds the roots of plants in the ground.

That way, the plants do not blow away.

Plant roots help prevent soil from washing away in the wind or rain.

People can use clay to create pots and many other objects.

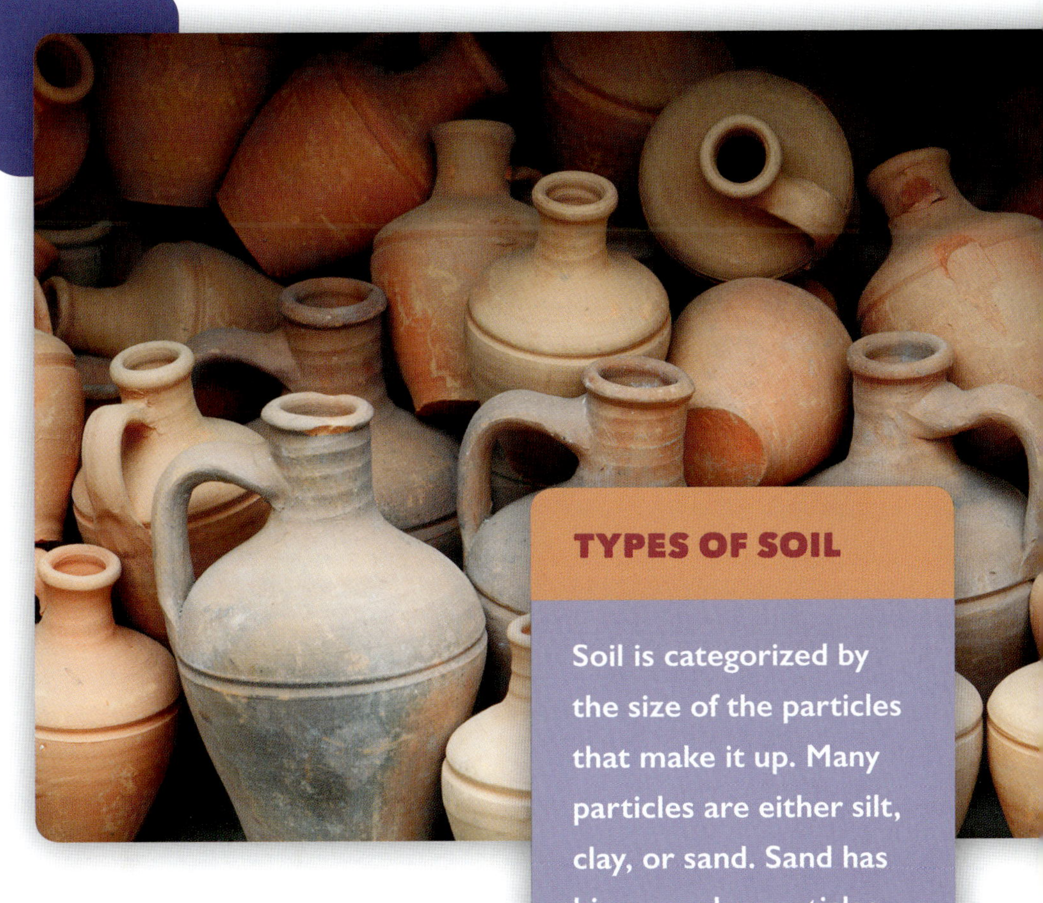

TYPES OF SOIL

Soil is categorized by the size of the particles that make it up. Many particles are either silt, clay, or sand. Sand has big, granular particles. Silt is made up of medium-sized particles. Clay is made from the smallest particles. Combinations of these particles are called loam.

Instead, they can use **nutrients** from the soil to grow.

Soil also acts as a filter. Water from the surface seeps down into the ground. The water moves through layers

The soil in a wetland helps filter the water that passes through it.

of soil. Some layers trap **bacteria** and **pollutants**. As a result, the water becomes cleaner and safer to drink.

CHAPTER 3
POSSIBLE PROBLEMS

One of the biggest dangers to soil is erosion. Erosion is when soil is removed by wind, water, or ice. Wind can blow soil away. Water can wash it to new places. Ice can scrape away layers of soil.

COMPLETE AN ACTIVITY HERE!

Moving water can wear away rock and soil.

DID YOU KNOW?

Erosion also wears away at rock. Wind, ice, and changes in temperature can break down rocks.

After erosion, there may not be enough soil to filter **pollutants** out of water. The polluted land does not hold water as well. This can lead to flooding. Flooding can lead to more erosion.

Heavy rain can cause flooding, which erodes riverbanks.

Dust storms are common in places that lack plants with deep root systems.

People can cause erosion too. They clear trees and plants to create farms. Or they feed plants to farm animals. Plants and trees help hold soil in place. Without them, soil is more likely to wash away or blow away.

Chemicals can also get into the soil. Farmers spray their crops with chemicals. Some can leak into the soil.

Chemicals can sometimes enter the soil through storm drains.

Factories also release chemicals into soil. These chemicals can make people or animals sick.

CHAPTER 4 HOW TO HELP

People can protect the soil in many ways. Planting trees along rivers and streams can help. So can planting grass. The roots keep the soil in place. It will not wash away as easily.

LEARN MORE HERE!

Soil's color can tell scientists a lot about its health.

Farmers can also help. Some leave crops in the ground after harvest. The dead plants add **nutrients** back to the soil. Some farmers plant crops in terraces. These ridges stop water

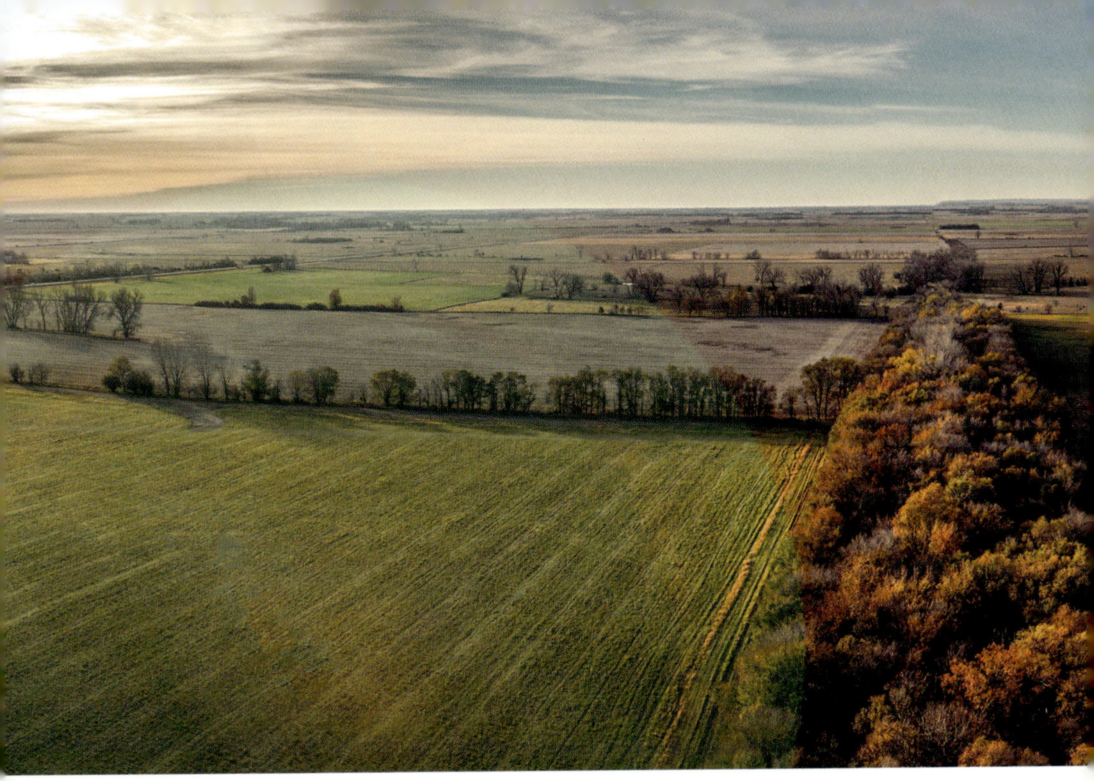

Wind erosion is a problem for farmers on prairies.

from running off the fields and taking soil with it.

DID YOU KNOW?

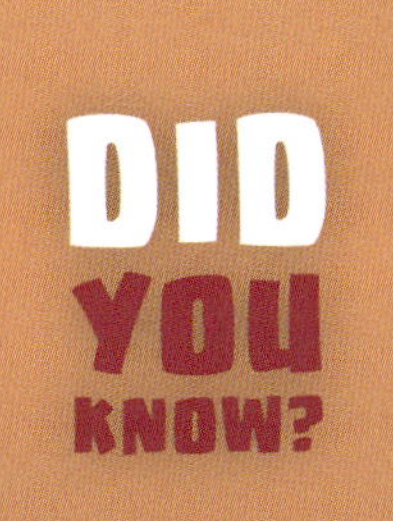

Some farmers plant windbreaks. These lines of trees go around a field. They block the wind and help prevent erosion.

Families can put fewer chemicals in their gardens and lawns. They can also use compost. To make compost, people collect food waste. They let it **decay**. Then they add it to the soil. The compost puts nutrients back into the soil. Actions like these help the soil stay healthy for many years.

Rich soil helps foods such as carrots and potatoes grow.

Vegetable peels, eggshells, and fruit waste make good compost.

PROTECTING SOIL IN THE UNITED STATES

1700s
George Washington and Thomas Jefferson begin experiments with ways to help soil.

1900s
The US government begins printing information about caring for soil.

1931
In the central United States, giant dust storms caused by erosion start an event known as the Dust Bowl.

1935
The Soil Conservation Service is created by the US government to teach landowners to care for soil.

1977
The US government passes a law to promote caring for soil and water.

2002
New York begins offering an award to farmers who keep soil safe on farmlands.

MAKING CONNECTIONS

TEXT-TO-SELF

Earth has many different kinds of soil. Where have you seen soil? What color or texture did it have?

TEXT-TO-TEXT

Have you read other books about soil? What is something new that you learned in this book?

TEXT-TO-WORLD

Plants need nutrients from the soil to grow. What could happen if all these nutrients get used up?

GLOSSARY

bacteria – tiny life-forms.

decay – to rot or break down.

mineral – a nonliving solid that forms naturally.

nutrient – matter that humans, animals, and plants need to stay strong and healthy.

organism – one of various kinds of living things, including plants, animals, and very small life-forms.

pollutant – something that harms water, soil, or air.

INDEX

Scan this code* and others like it while you read, or visit the website below to make this book pop!

popbooksonline.com/soil

*Scanning QR codes requires a web-enabled smart device with a QR code reader app and a camera.